MATERIAL BALANCE CALCULATIONS (REVISED)

A Step-By-Step Explanation

With Numerous Worked Examples

By

Kingsley Augustine

TABLE OF CONTENT

CHAPTER 1
MATERIAL BALANCE: INTRODUCTION

A material balance is an expression indicating the law of conservation of matter which account for all the materials in a process.

It can be expressed as:

Input - Output = Accumulation

In a process where a chemical reaction occurs, the material balance equation becomes:

Input - Output = Generation by reaction + Accumulation

When the process is a steady state process where there is no accumulation, then the material balance expression is given by:

Input - Output = Generation by reaction

Note that when a component goes into a process and does not undergo any change, then:

Input = Output

This is a case where there is no accumulation and no reaction.

Calculations in material balances are usually carried out in masses, which also involves mass flow rate. The use of mass fraction is also very important. Volumetric flow rate and mole fraction are also useful for material balance calculations.

CHAPTER 2
BALANCES INVOLVING DRYING/EVAPORATIVE PROCESSES

In processes involving drying or evaporation, one of the components usually passes the process without undergoing any change. Only water is removed from the wet solid.

Examples

1. 800kg of wet flour is fed into a dryer. The wet flour contains 40% by mass of water, while the dry product is to contain 5% by mass of water. Calculate the mass of water removed by the dryer.

Solution

Let F = mass of wet flour (i.e. feed)

   W = mass of water removed

and P = mass of product (dried flour)

Also, let x_F = % of water wet flour (mass fraction of water in feed)

      x_W = % of water in water (this must be 1)

and x_P = % of water in product

Therefore, the overall component balance is given by:

   Input = Output   (Since there is no accumulation)

   $F = W + P$ ....Equation 1

The water component balance is given by:

Water in wet flour = water removed + water in product

Therefore, $Fx_F = Wx_W + Px_P$ ....Equation 2

Note that amount of water in a substance = % of water in the substance x mass of the substance

Therefore from equation 1,

   $F = W + P$

800 = W + P ....Equation 3

From equation 2:

$Fx_F = Wx_W + Px_P$ (Note that $x_W = 1$, since it contains only water, $x_F = \dfrac{40}{100} = 0.4$, and $x_P = \dfrac{5}{100} = 0.05$). Hence:

800 x 0.4 = (W x 1) + P x 0.05

320 = W + 0.05P ....Equation 4

From equation 3:

W = 800 - P ....Equation 5

Substitute 800 - P for W into equation 4. This gives:

320 = W + 0.05P ....Equation 4

320 = (800 - P) + 0.05P

320 = 800 - 0.95P   (Note that P - 0.05P = 0.95P)

0.95P = 800 - 320

0.95P = 480

$P = \dfrac{480}{0.95}$

P = 505.3kg

From equation 5:

W = 800 - P

= 800 - 505.3

W = 294.7kg

Therefore the mass of water removed is 294.7kg.

This process can be summarized as follows:

In the feed: mass of water = 0.4 x 800 = 320kg

mass of flour = 800 = 320 = 480kg

In the product: mass of water in product = 0.05 x 505.3 = 25.3kg

mass of flour = 505.3 - 25.3

=480kg (This shows that the flour has not changed from feed)

Mass of water removed = 294.7kg

Note that x_F was converted to fraction as: $\frac{40}{100}$ = 0.4. Similarly, $x_P = \frac{5}{100}$ = 0.05. The percentage of water in water, x_W = 100%, which was converted to fraction as: $\frac{100}{100}$ = 1

2. A paper pulp containing 60% of moisture is passed through a dryer in order to remove 80% of the original moisture. Calculate:

(a). The mass of water removed per kg of the pulp

(b). The composition of the product

Solution

Let F = paper pulp (i.e. the feed)

W = moisture removed

P = product (i.e. dried pulp)

Also, let x_F = % moisture in feed = 60% = 0.6

x_W = % moisture in moisture = 100% (water only) = 1

x_P = % moisture in product

The overall component balance is given by:

F = W + P

1 = W + P ....Equation 1   (Note that F = 1kg since we are to calculate for per kg of pulp)

The moisture component balance is given by:

$Fx_F = Wx_W + Px_P$

6

$1 \times 0.6 = (W \times 1) + Px_P$

$0.6 = W + Px_P$ ....Equation 2

From equation 1:

$P = 1 - W$ ....Equation 3

Amount of moisture in feed = 60%

Therefore moisture in feed = 0.6 x 1 = 0.6kg

Since 80% of this moisture was removed, then:

Amount of moisture removed = 0.8 x 0.6 = 0.48kg

Therefore, W = 0.48kg

(b). From equation 3,

$P = 1 - W$

$= 1 - 0.48$

$P = 0.52kg$

Substituting the values of W and P into equation 2 gives:

$0.6 = W + Px_P$

$0.6 = 0.48 + 0.52x_P$

$0.6 - 0.48 = 0.52x_P$

$0.52x_P = 0.12$

$x_P = \dfrac{0.12}{0.52}$

$x_P = 0.23$

Recall that the amount of moisture in feed = 0.6kg.

Therefore, the amount of pulp in feed = 1 - 0.6 = 0.4kg

This amount of pulp in feed, also passes completely to the product. Therefore, amount of pulp in product is 0.4kg.

But, product, P = 0.52kg

Therefore amount of water in product, i.e. dried pulp = 0.52 - 0.4

$$= 0.12kg.$$

Therefore the composition of the product is:

$$\% \text{ of pulp} = \frac{0.4}{0.52} \times 100 = 76.9\%$$

$$\% \text{ of moisture} = \frac{0.12}{0.52} \times 100 = 23.1\%$$

3. $50m^3/h$ of a 5% sodium chloride solution of density $1200kg/m^3$ is concentrated by evaporation to produce 40% brine. Calculate the rate of evaporation of water from the salt solution.

Solution

Let F = mass flow rate of the feed, in kg/h

W = water removed in kg/h

P = product i.e. brine in kg/h

Also, let, x_F = % salt in feed = 0.05   (Salt is the NaCl)

x_W = % salt in water = 0 (No salt in evaporated water)

x_P = % salt in product = 0.4

Let us convert the volumetric flow rate to mass flow rate

Recall that: Density $= \dfrac{mass}{volume}$

Therefore mass = density x volume

Similarly, mass flow rate = density x volumetric flow rate

$$= 1200 \times 50$$

$$= 60,000kg/h$$

Therefore, F = 60,000kg/h

The overall component balance is given by:

$$F = W + P$$

$$60{,}000 = W + P \dots\dots\dots\dots\text{Equation 1}$$

The salt component balance is given by:

$$Fx_F = Wx_W + Px_P$$

$$60{,}000 \times 0.05 = (W \times 0) + (P \times 0.4)$$

$$3000 = 0.4P$$

Therefore, $P = \dfrac{3000}{0.4}$

$$P = 7500\text{kg/h}$$

From equation 1:

$$60{,}000 = W + P$$

$$60{,}000 = W + 7500$$

$$W = 60{,}000 - 7500$$

$$W = 52{,}500\text{kg/h}$$

Therefore, the rate of evaporation of water is 52,500kg/h.

4. A fresh wood containing 45% of moisture is dried in an oven in order to remove 95% of its moisture content. Calculate:

(a). The mass of water remove per 100kg of the wood

(b). The mass of the dried wood

Solution

Let F = fresh wood (i.e. the feed)

W = moisture removed

P = product (i.e. dried pulp)

Also, let x_F = % moisture in feed = 45% = 0.45

x_W = % moisture in moisture = 100% (water only) = 1

x_P = % moisture in product

The overall component balance is given by:

$F = W + P$

$100 = W + P$ ....Equation 1   (Note that F = 100kg)

The moisture component balance is given by:

$Fx_F = Wx_W + Px_P$

$100 \times 0.45 = (W \times 1) + Px_P$

$45 = W + Px_P$ ....Equation 2

From equation 1:

$P = 100 - W$ ....Equation 3

Amount of moisture in feed = 45%

Therefore moisture in feed = 0.45 x 100 = 45kg

Since 95% of this moisture was removed, then:

Amount of moisture removed = 0.95 x 45

= 42.75kg

Therefore, W = 42.75kg

(b). From equation 3,

$P = 100 - W$

$= 100 - 42.75$

$P = 57.25kg$

Therefore the mass of the dried wood is 57.25kg

5. 2000kg of a 12% sugar solution is concentrated by evaporation to produce a 70% sugar solution. Calculate the mass of water evaporated.

Solution

Let F = mass of the feed

    W = water removed

    P = product

Also, let, x_F = % sugar in feed = 0.12

    x_W = % sugar in water = 0 (No sugar in evaporated water)

    x_P = % sugar in product = 0.7

The overall component balance is given by:

    F = W + P

    2000 = W + P ....Equation 1

The sugar component balance is given by:

    $Fx_F = Wx_W + Px_P$

    2000 x 0.12 = (W x 0) + (P x 0.7)

    240 = 0.7P

Therefore, $P = \dfrac{240}{0.7}$

    P = 342.9kg

From equation 1:

    2000 = W + P

    2000 = W + 342.9

    W = 2000 - 342.9

    W = 1657.1kg

Therefore, the mass of water evaporated is 1657.1kg.

EXERCISE

1. 2500kg of wet flour is fed into a dryer. The wet flour contains 45% by mass of water, while the dry product is to contain 4% by mass of water. Calculate the mass of water removed by the dryer.

2. A fabric material containing 72% of moisture is passed through a dryer in order to remove 94% of the original moisture. Calculate:

(a). The mass of water removed per kg of the fabric

(b). The composition of the product

3. 200m^3/h of a 7% sodium chloride solution of density 1910kg/m^3 is concentrated by evaporation to produce 85% brine. Calculate the rate of evaporation of water from the salt solution.

4. A fresh wood containing 82% of moisture is dried in an oven in order to remove 88% of its moisture content. Calculate:

(a). The mass of water remove per 5kg of the wood

(b). The mass of the dried wood

5. 400kg of a 5% sugar solution is concentrated by evaporation to produce a 58% sugar solution. Calculate the mass of water evaporated.

CHAPTER 3
BALANCES INVOLVING MIXING OF SOLUTIONS

In mixing of solutions, solutions are usually fed into a process to produce one output, provided the solutions do not react together.

Examples

1. 14% by mass of tetraoxosulphate (VI) acid solution is added into a process tank. 210kg of 75% tetraoxosulphate (VI) acid solution is also added to the tank. If the final solution gives 30% of sulphuric acid, determine the mass of this final acid solution.

Solution

Let F_1 = 14% tetraoxosulphate (VI) acid, i.e. the dilute acid

F_2 = 75% tetraoxosulphate (VI) acid, i.e. the concentrated acid

P = Product   (The final acid solution)

Also, let x_{F1} = % by mass of acid in F_1 = 0.14

x_{F2} = % by mass of acid in F_2 = 0.75

x_P = % by mass of acid in P = 0.3

Therefore, the overall component balance is given by:

$F_1 + F_2 = P$

$F_1 + 210 = P$ ....Equation 1

The tetraoxosulphate (VI) acid component balance is given by:

$F_1x_{F1} + F_2x_{F2} = Px_P$

$(F_1 \times 0.14) + (210 \times 0.75) = (P \times 0.3)$

$0.14F_1 + 157.5 = 0.3P$ ....Equation 2

From equation 1:

$F_1 = P - 210$

Substitute P - 210 for F_1 in equation 2. This gives:

$$0.14F_1 + 157.5 = 0.3P \ldots\ldots\ldots\ldots\ldots\text{Equation 2}$$

$$0.14(P - 210) + 157.5 = 0.3P$$

$$0.14P - 29.4 + 157.5 = 0.3P$$

$$157.5 - 29.4 = 0.3P - 0.14P$$

$$128.1 = 0.16P$$

$$P = \frac{128.1}{0.16}$$

$$P = 800.6kg$$

Therefore the mass of this final acid solution is 800.6kg.

Note that the mass of tetraoxosulphate (VI) acid in this final solution is $= Px_P = 800.6 \times 0.3 = 240.2kg$

2. A fuel oil seller mixes two types of oils each containing n-heptane and iso-octane. He decides to mix 800kg of 90% iso-octane which is the first oil, with 70% iso-octane which is the second oil. If this produces 85% iso-octane, calculate:

(a). The mass of the second oil added to the mixture

(b). The mass of iso-octane in the product

Solution

Let F_1 = mass of first oil

F_2 = mass of second oil

P = mass of product

Also let x_{F1} = % by mass of iso-octane in F_1 = 0.9

x_{F2} = % by mass of iso-octane in F_2 = 0.7

x_P = % by mass of iso-octane in P = 0.85

Therefore the overall component mass balance is given by:

$$F_1 + F_2 = P$$

$800 + F_2 = P$ ....Equation 1

The iso-octane component balance is given by:

$F_1 x_{F1} + F_2 x_{F2} = P x_P$

$(800 \times 0.9) + (F_2 \times 0.7) = P \times 0.85$

$720 + 0.7 F_2 = 0.85P$ ....Equation 2

From equation 1:

$P = 800 + F_2$ ....Equation 3

Substituting $800 + F_2$ for P in equation 2 gives:

$720 + 0.7 F_2 = 0.85P$ ....Equation 2

$720 + 0.7 F_2 = 0.85(800 + F_2)$

$720 + 0.7 F_2 = 680 + 0.85 F_2$

$720 - 680 = 0.85 F_2 - 0.7 F_2$

$40 = 0.15 F_2$

$F_2 = \dfrac{40}{0.15}$

$F_2 = 266.7$

The mass of the second oil in the mixture 266.7kg

(b). From equation 3 above:

$P = 800 + F_2$ ....Equation 3

$= 800 + 266.7$

$P = 1066.7kg$

But mass of iso-octane in the product is given by:

$P x_P$

$= 1066.7 \times 0.85$

$= 906.7kg$

15

Therefore mass of iso-octane in the product is 906.7kg.

3. A fruit juice is made from a mixture of juice and water. A fruit juice containing 20% of juice and 80% of water sells for $90/kg. It is mixed with a fruit juice containing 40% of juice which sells for $50/kg. If the fruit juice produced contains 34% of juice, how much should it be sold to make a profit of 20%

Solution

The prices of the fruit juice are the cost of 1kg of the fruit juice.

So, let us solve this problem by taking a basis of 1kg of product produced.

Let F_1 = mass of 20% juice

F_2 = mass of 40% juice

P = mass of product = 1kg

Also, let x_1 = % by mass of juice in F_1 = 0.20

x_2 = % by mass of juice in F_2 = 0.4

x_P = % by mass of juice in P = 0.34

The overall component balance is given by:

$F_1 + F_2 = P$

$F_1 + F_2 = 1$ (Since P = 1kg)

Or, $F_1 = 1 - F_2$ ....Equation 1

The juice component balance is given by:

$F_1x_1 + F_2x_2 = Px_P$

$0.20F_1 + 0.4F_2 = 1 \times 0.34$

$0.2F_1 + 0.4F_2 = 0.34$ ....Equation 2

Substitute 1 - F_2 for F_1 from equation 1 into equation 2. This gives:

$0.2F_1 + 0.4F_2 = 0.34$ ....Equation 2

$0.2(1 - F_2) + 0.4F_2 = 0.34$

$0.2 - 0.2F_2 + 0.4F_2 = 0.34$

$0.4F_2 - 0.2F_2 = 0.34 - 0.2$

$0.2F_2 = 0.14$

$$F_2 = \frac{0.14}{0.2}$$

$F_2 = 0.7kg$

From equation 1:

$F_1 = 1 - F_2$

$= 1 - 0.7$

$F_1 = 0.3kg$

From the question, 1 kg of F_1 cost \$90 while 1kg of F_2 cost \$50. Therefore, 0.3kg of F_1 and 0.7kg of F_2 will cost:

$(0.3 \times 90) + (0.7 \times 50)$

$= 27 + 35$

$= \$62$

In order to make 20% profit on this cost, the juice must be sold for:

120% of cost price   [since 20% profit means 120% (i.e. 100 + 20 = 120) of cost price]

$$= \frac{120}{100} \times 62$$

$= 74.4$

The fruit juice should be sold for \$74.40

4. 400 litres of 80% by mass of HCl is to be obtained by mixing two solutions of HCl. If this is obtained by mixing 90% of HCl and 50% of HCl solutions, calculate the volume (in litres) of each of these two solutions that are mixed together.

(Density of 80% HCl = 1.34kg/litre, Density of 90% HCl = 1.42kg/litre, Density of 50% HCl = 1.25kg/litre)

<u>Solution</u>

Since mass fractions are given, the volume of 400 litres must be converted to mass.

Recall that, density = $\dfrac{mass}{volume}$

Therefore, mass = density x volume

Hence, mass of 400 litres, 80% HCl = 1.34 x 400   (Its density is 1.34)

$$= 536kg$$

Hence, the product, P = 536kg

Let: F_1 = 90% HCl

   F_2 = 50% HCl

   x_1 = 0.9

   x_2 = 0.5

   x_P = 0.8   (Note that the 400 litres 80% HCl is the product)

The overall component mass balance is given by:

   $F_1 + F_2 = P$

   $F_1 + F_2 = 536$ ....Equation 1

The HCl component balance is given by:

   $F_1x_1 + F_2x_2 = Px_P$

(F_1 x 0.9) + (F_2 x 0.5) = 536 x 0.8

       $0.9F_1 + 0.5F_2 = 428.8$ ....Equation 2

From equation 1:

       $F_1 = 536 - F_2$ ....Equation 3

Substitute 536 - F_2 for F_1 in equation 2

$0.9F_1 + 0.5F_2 = 428.8$ ....Equation 2

$0.9(536 - F_2) + 0.5F_2 = 428.8$

$482.4 - 0.9F_2 + 0.5F_2 = 428.8$

$482.4 - 428.8 = 0.9F_2 - 0.5F_2$

$53.6 = 0.4F_2$

$$F_2 = \frac{53.6}{0.4}$$

$F_2 = 134kg$

From equation 3:

$F_1 = 536 - F_2$ ....Equation 3

$= 536 - 134$

$F_1 = 402kg$

In order to obtain F_1 and F_2 in litres (volume), their densities have to be applied.

Recall that, density $= \dfrac{mass}{volume}$

Therefore, volume $= \dfrac{mass}{density}$

Hence, volume of F_1 in litres $= \dfrac{402}{1.42}$ (F_1 is 90% HCl and has density of 1.42)

$= 283.8$ litres

Volume of F_2 in litres $= \dfrac{134}{1.25}$ (F_2 = 50% HCl and has a density of 1.25)

$= 107.2$ litres

5. A paint seller mixes two types of paints. A paint containing 30% of pigment sells for $20/kg. It is mixed with a paint containing 65% of pigment which sells for $52/kg. If the paint produced contains 46% of pigment, how much should it be sold?

Solution

The prices of the paint are the cost of 1kg of the paint.

So, let us solve this problem by taking 1kg of product formed.

Let F_1 = mass of 30% pigment

F_2 = mass of 65% pigment

P = mass of product = 1kg

Also, let x_1 = % by mass of pigment in F_1 = 0.30

x_2 = % by mass of pigment in F_2 = 0.65

x_P = % by mass of pigment in P = 0.46

The overall component balance is given by:

$F_1 + F_2 = P$

$F_1 + F_2 = 1$ (Since P = 1kg)

Or, $F_1 = 1 - F_2$ ....Equation 1

The pigment component balance is given by:

$F_1x_1 + F_2x_2 = Px_P$

$0.30F_1 + 0.65F_2 = 1 \times 0.46$

$0.3F_1 + 0.65F_2 = 0.46$ ....Equation 2

Substitute 1 - F_2 for F_1 from equation 1 into equation 2. This gives:

$0.3F_1 + 0.65F_2 = 0.46$ ....Equation 2

$0.3(1 - F_2) + 0.65F_2 = 0.46$

$0.3 - 0.3F_2 + 0.65F_2 = 0.46$

$0.65F_2 - 0.3F_2 = 0.46 - 0.3$

$0.35F_2 = 0.16$

$$F_2 = \frac{0.16}{0.35}$$

$F_2 = 0.457kg$

From equation 1:

$F_1 = 1 - F_2$

$= 1 - 0.457$

$F_1 = 0.543kg$

From the question, 1 kg of F_1 cost $20 while 1kg of F_2 cost $52. Therefore, 0.543kg of F_1 and 0.457kg of F_2 will cost:

$(0.543 \times 20) + (0.457 \times 52)$

$= 10.86 + 23.76$

$= \$34.62$

Hence the paint should be sold for $34.62

EXERCISE

1. 10% by mass of tetraoxosulphate (VI) acid solution is added into a process tank. 500kg of 50% tetraoxosulphate (VI) acid solution is also added to the tank. If the final solution gives 24% of sulphuric acid, determine the mass of this final acid solution.

2. A fuel oil seller mixes two types of oils each containing n-heptane and iso-octane. He decides to mix 1500kg of 80% iso-octane which is the first oil, with 35% iso-octane which is the second oil. If this produces 66% iso-octane, calculate:

(a). The mass of the second oil added to the mixture

(b). The mass of iso-octane in the product

3. A fruit juice is made from a mixture of juice and water. A fruit juice containing 18% of juice and 82% of water sells for $60/kg. It is mixed with a fruit juice containing 45% of juice which sells for $34/kg. If the fruit juice produced contains 40% of juice, how much should it be sold to make a profit of 45%

4. 20 litres of 90% by mass of HCl is to be obtained by mixing two solutions of HCl. If this is obtained by mixing 98% of HCl and 64% of HCl solutions, calculate the volume (in litres) of each of these two solutions that are mixed together.

(Density of 90% HCl = 1.15kg/litre, Density of 98% HCl = 1.28kg/litre, Density of 64% HCl = 1.10kg/litre)

5. A paint seller mixes two types of paints. A paint containing 50% of pigment sells for $12/kg. It is mixed with a paint containing 22% of pigment which sells for $5/kg. If the paint produced contains 36% of pigment, how much should it be sold?

CHAPTER 4
BALANCES ON SEPARATION PROCESSES

Material balances can be carried out on separation processes such as distillation processes, crystallization and filtration processes.

Examples

1. A distillation column is used to separate 1000kg/h of a mixture containing 60% water and 40% ethanol. The product at the top of the column contains 85% ethanol, while the product at the bottom contains 90% water. The vapour enters the condenser from the top of the column at 640kg/h and is returned to column as reflux. The rest is withdrawn. Calculate:

(a). the rate of formation of the top product and the bottom product

(b). the flow rate of obtaining ethanol in the top product

(c). the rate of production of ethanol in the bottom product

(d). the ratio of the amount refluxed to the product withdrawn

Solution

(a). Let us use a basis of 1000kg/h of the feed.

Let, F = feed rate = 1000kg/h

    D = distillate rate (top product)

    W = bottom product rate

    x_F = mass fraction of ethanol in feed = 0.4

    x_D = mass fraction of ethanol in the top product = 0.85

    x_W = mass fraction of ethanol in the bottom product = 1 - 0.9 = 0.1 (water = 0.9)

The overall mass component balance is given by:

    F = D + W   (since input = output)

1000 = D + W ....equation 1

The ethanol component balance is given by:

$$Fx_F = Dx_D + Wx_W$$

$$1000(0.4) = D(0.85) + W(0.1)$$

$$400 = 0.85D + 0.1W \text{ ....Equation 2}$$

From equation 1: $D = 1000 - W$ ....Equation 3

Substitute 1000 - W for D in equation 2

$$400 = 0.85D + 0.1W \text{ ....Equation 2}$$

$$400 = 0.85(1000 - W) + 0.1W$$

$$400 = 850 - 0.85W + 0.1W$$

$$400 = 850 - 0.75W$$

$$0.75W = 850 - 400$$

$$0.75W = 450$$

$$W = \frac{450}{0.75}$$

$$W = 600$$

From equation 3:

$$D = 1000 - W$$

$$= 1000 - 600$$

$$D = 400$$

Therefore the top product rate is 400kg/h, while the bottom product rate is 600kg/h

(b) The flow rate of obtaining ethanol in the top product = Dx_D

$$= 400 \times 0.85$$

$$= 340kg/h$$

(c). The rate of producing ethanol in the bottom product = Wx_W

$$= 600 \times 0.1$$

$$= 60kg/h$$

(d). The balance about the condenser is given by:

$$V = R + D$$

where V = vapour rate to the column, R = reflux rate, D = distillate rate (top product)

Therefore, V = R + D

$$640 = R + 400$$

$$R = 640 - 400$$

$$R = 240$$

Therefore the ratio of the amount refluxed to the product withdrawn which is the reflux ratio is given by:

$$\text{Reflux ratio} = \frac{R}{D}$$

$$= \frac{240}{400}$$

$$= 0.6$$

2. An alcohol-water mixture contains 50% by mass of alcohol. A distillation column is used to separate the mixture by a continuous process. If the composition of water in the top product is 15%, and it is 80% at the bottom, determine:

(a). the amount of alcohol in the top product

(b). what percentage of the feed is the top product

(c). the amount of alcohol in the top product as a percentage of alcohol in the feed

(d). the amount of water in the bottom product.

Solution

(a). The overall component balance on a basis of 100kg/h feed rate is given by:

$$F = D + W$$

$$100 = D + W \dots\dots\dots\dots\dots\dots\text{Equation 1}$$

The alcohol component balance is given by:

$$Fx_F = Dx_D + Wx_W$$

But $x_F = 0.5$, $x_D = 1 - 0.15 = 0.85$, and $x_W = 1 - 0.8 = 0.2$

Therefore, $Fx_F = Dx_D + Wx_W$

$$100(0.5) = D(0.85) + W(0.2)$$

$$50 = 0.85D + 0.2W \quad ....Equation\ 2$$

From equation 1: $D = 100 - W$ ....Equation 3

Substitute 100 - W for D in equation 2

$$50 = 0.85D + 0.2W \quad ....Equation\ 2$$

$$50 = 0.85(100 - W) + 0.2W$$

$$50 = 85 - 0.85W + 0.2W$$

$$0.85W - 0.2W = 85 - 50$$

$$0.65W = 35$$

$$W = 35/0.65$$

$$W = 53.8$$

From equation 3, $D = 100 - W$

$$= 100 - 53.8$$

$$D = 46.2$$

Top product, D = 46.2kg/h, and bottom product, W = 53.8kg/h

Amount of alcohol in the top product is = Dx_D

$$= 46.2 \times 0.85$$

$$= 39.3kg/h$$

(b). Feed = 100kg/h, top product = 46.2kg/h

Therefore percent of feed which is the top product = $\dfrac{46.2}{100} \times 100$

$$= 46.2\%$$

(c) Alcohol in feed = Fx_F

$$= 100 \times 0.5$$

$$= 50kg/h.$$

Alcohol in the top product = 39.3kg/h

Therefore amount of alcohol in the top product as a percentage of alcohol in the feed is:

$$= \frac{39.3}{50} \times 100$$

$$= 78.6\%$$

(d) The amount of water in the bottom product = W x mass fraction of water in the bottom product = 53.8 x 0.8

$$= 43.0kg/h$$

3. Water and air flow into a humidifier in which the water evaporates completely into the air. The entering air contains 1.5mole% water vapour. The outlet humidified air contains 20mole% water. Calculate the volume of water in m^3/min required to humidify 100mole/min of the entering air.

Solution

Let F_1 = moles/min of inlet water

F_2 = moles/min of inlet air

P = moles/min of product (humidified air)

Also, let m_1 = mole fraction of water in inlet water = 1 (since inlet water contains only water)

m_2 = mole fraction of water in inlet air = 0.015 (from 1.5% in the question)

m_3 = mole fraction of water in product = 0.2 (from 20%)

The overall component balance is given by:

$$F_1 + F_2 = P$$

$$F_1 + 100 = P \quad ....\text{Equation 1}$$

The water component balance is given by:

$F_1 m_1 + F_2 m_2 = P m_3$

$F_1(1) + 100(0.015) = P(0.2)$

$F_1 + 1.5 = 0.2P$ ....Equation 2

Substitute $F_1 + 100$ (equation1) for P into equation 2. This gives:

$F_1 + 1.5 = 0.2P$ ....Equation 2

$F_1 + 1.5 = 0.2(F1 + 100)$

$F_1 + 1.5 = 0.2F1 + 20$

$F_1 - 0.2F_1 = 20 - 1.5$

$0.8F_1 = 18.5$

Therefore, $F_1 = \dfrac{18.5}{0.8}$

$F_1 = 23.1 \text{mole/min}$

Recall that 1 mole of water = 18g   (i.e. the molecular mass of water)

Therefore, 23.1moles 0f water = 23.1 x 18

$= 415.8g$

Therefore mass of inlet water = 415.8g/min

Recall that: Density $= \dfrac{mass}{volume}$

$1g/cm^3 = \dfrac{mass \ in \ g}{volume \ in \ cm^3}$

Therefore volume in $cm^3 = \dfrac{mass}{1}$

$= \dfrac{415.8}{1}$

$= 415.8 cm^3/min$

Let us now convert this volume to m^3/min. In order to convert cm^3 to m^3, we divide by 100^3.

Therefore the volume in m^3/min $= \dfrac{415.8}{100^3}$

$$= \frac{415.8}{10^6} \quad (100^3 = 100 \times 100 \times 100 = 10^6)$$

$$= 4.158 \times 10^{-4} \text{m}^3/\text{min}$$

4. An air conditioner takes in 250m^3/min of air at 25°C, 720mmHg pressure and 92% relative humidity, and gives out the air at 12°C. Calculate:

(a). the flow rate of the outlet air

(b). the rate at which condensed water must be removed from the air conditioner.

(Take saturated vapour pressure of water at 25°C as 23.7mmHg, and at 12°C as 10.5mmHg)

Solutions

Let us calculate the mole fraction water in the inlet air. This can be obtained by using the formula:

$$y_1 = \frac{H_R}{100} \times \frac{P_t}{P_D}$$

where y_1 = mole fraction, H_R = relative humidity, P_t = saturated vapour pressure of water at air temperature, and P_D = saturated vapour pressure of water at dew point.

Therefore, $y_1 = \frac{H_R}{100} \times \frac{P_t}{P_D}$

$$= \frac{92}{100} \times \frac{23.7}{720}$$

$$y_1 = 0.0303$$

Therefore mole fraction of the dry air = 1 - 0.0303

$$= 0.9697$$

Let us also calculate the molar flow rate of the inlet air. In order to do this, recall the ideal gas equation: PV = mRT

Therefore, the standard condition can be expressed as:

$$P_1 V_1 = m_1 RT \text{ ....Equation 1}$$

where P_1 = 760mmHg, V_1 = 22.4m^3 (molar volume of gas), m_1 = 1kmol (since 1kmol = 22.4m^3), and T_1 = 273k

The inlet condition of the air can be expressed as:

$$P_2V_2 = m_2RT \quad ....\text{Equation 2}$$

where P_2 = 720mmHg, V_2 = 250m^3/min, T_2 = 273 + 25 = 298k, m_2 = ?

Dividing equation 1 by equation 2 gives:

$$\frac{P_1V_1}{P_2V_2} = \frac{m_1T_1}{m_2T_2} \quad \text{(R has cancelled out)}$$

This gives an expression relating initial and final conditions as follows:

$$\frac{P_1V_1m_2}{T_1} = \frac{P_2V_2m_1}{T_2}$$

Therefore, $m_2 = \dfrac{P_2V_2T_1m_1}{P_1V_1T_2}$

$$= \frac{720 \times 250 \times 1 \times 273}{760 \times 22.4 \times 298}$$

$$m_2 = 9.69\text{kmols/min}$$

At the outlet condition, the mole fraction of water in the air can be obtained by using Raoult's law as follows:

$$y_2P_D = P_t \quad (P_D = 720\text{mmHg}, P_t = 10.5)$$

Therefore, $y_2 = \dfrac{P_t}{P_D}$

$$= \frac{10.5}{720}$$

$$= 0.0146$$

Therefore mole fraction of dry outlet air = $1 - y_2$

$$= 1 - 0.0146$$

$$= 0.9854$$

The dry air component balance is given by:

Input = Output

$$(1 - y_1)m_2 = (1 - y_2)m_3$$

where m_3 is the kmol/min of the outlet air.

Therefore, $0.9697 \times 9.69 = 0.9854 \times m_3$

Hence, $m_3 = \dfrac{0.9697 \times 9.69}{0.9854}$

$\qquad m_3 = 9.54 \text{kmols/min}$

Using this m_3, the outlet volume, V_3, can be calculated by using the expression below.

$$\frac{P_1 V_1 m_3}{T_1} = \frac{P_3 V_3 m_1}{T_3}$$

Therefore, $V_3 = \dfrac{P_1 V_1 m_3 T_3}{P_3 m_1 T_1}$

$$= \frac{760 \times 22.4 \times 9.54 \times (273 + 12)}{720 \times 1 \times 273} \qquad \text{(Note that } T_3 = 273 + 12\text{)}$$

$$= \frac{760 \times 22.4 \times 9.54 \times 285}{720 \times 273}$$

$\qquad V_3 = 235.5 \text{m}^3/\text{min}$

(b). The water component balance is given by:

$\qquad m_2 y_1 = m_3 y_2 + m_4 \qquad$ (where m_4 = kmol/min outlet water)

$\qquad 9.69 \times 0.0303 = (9.54 \times 0.0146) + m_4$

$\qquad 0.2936 = 0.1393 + m_4$

$\qquad m_4 = 0.2936 - 0.1393$

$\qquad m_4 = 0.154$

Therefore the condensed water must leave at a rate of 0.154kmols/min.

5. A gas containing vapour enters into a condenser. The partial pressure of the vapour in the gas is 800mmHg. The partial pressure of the vapour in the outlet gas is 780mmHg at a temperature of 140°C. If the total pressure of the system is 860mmHg, calculate the volume of the gas leaving the condenser when 100kmols of the vapour condenses out.

Solution

From Raoult's law:

$$p = yP$$

Or, $y = \dfrac{p}{P}$

where p = partial pressure, and P = total pressure. At the inlet condition, the mole fraction of the vapour, y_1, is obtained as follows:

$$y_1 = \dfrac{p}{P}$$

$$= \dfrac{800}{860}$$

$$y_1 = 0.930$$

At the outlet condition, $y_2 = \dfrac{780}{860}$

$$y_2 = 0.907$$

The overall component balance is given by:

$$m_1 = m_2 + m_3$$

where m_1 = inlet gas, m_2 = outlet dry gas, and m_3 = outlet vapour = 100

Hence, $m_1 = m_2 + 100$ ....Equation 1

The vapour component balance is given by:

$$m_1 y_1 = m_2 y_2 + m_3 y_3 \quad (y_3 = 1, \text{ since only vapour is in } m_3)$$

Hence. $m_1 y_1 = m_2 y_2 + (100 \times 1)$

$$m_1 y_1 = m_2 y_2 + 100 \text{ ....Equation 2}$$

Substitute $m_2 + 100$ (in equation 1) for m_1 into equation 2. This gives:

$$m_1 y_1 = m_2 y_2 + 100 \text{ ....Equation 2}$$

$$(m_2 + 100)0.930 = m_2(0.907) + 100$$

$$0.93 m_2 + 93 = 0.907 m_2 + 100$$

$$0.93 m_2 - 0.907 m_2 = 100 - 93$$

$$0.023m_2 = 7$$

$$m_2 = \frac{7}{0.023}$$

$$m_2 = 304.3 \text{kmols}$$

Therefore 304.3kmols of the dry gas leaves the condenser. This can be converted to volume by using the expression below.

$$\frac{P_1 V_1 m_2}{T_1} = \frac{P_2 V_2 m_1}{T_2}$$

where P_1 = 760mmHg, V_1 = 22.4m^3, m_1 = 1kmol, T_1 = 273k, while the other quantities have their usual meanings.

Therefore, $\quad V_2 = \dfrac{P_1 V_1 m_2 T_2}{P_2 m_1 T_1}$

$$= \frac{760 \times 22.4 \times 304.3 \times (273 + 140)}{860 \times 273}$$

$$= \frac{760 \times 22.4 \times 304.3 \times 413}{860 \times 273}$$

$$= 9113 \text{m}^3$$

Therefore volume of the gas leaving the condenser is 9113m^3

6. Wet paper pulp with a moisture content of 72% enters a continuous process dryer and leaves at a rate of 50kg/h with a moisture content of 8%. Dry air at 66°C and 76cmHg enters the dryer. The air and water vapour leaves the dryer at 55°C and 76cmHg. Calculate:

(a). the rate at which the wet pulp enters the dryer

(b). the flow rate of the inlet air in m^3/h if the outlet air has a relative humidity of 45%

Solution

(a). Let us work with the rate of 50kg/h dried pulp. The dry pulp component balance is given by:

Input = output

$$m_2(1 - x_2) = (1 - x_4)m_4$$

where m_2 = inlet wet pulp flow rate in kg/h

m_4 = outlet pulp flow rate in kg/h = 50kg/h

x_2 = mass fraction of moisture in the inlet pulp = 0.72

x_4 = mass fraction of moisture in the outlet pulp = 0.08

Hence, $m_2(1 - x_2) = (1 - x_4)m_4$

$m_2(1 - 0.72) = (1 - 0.08)m_4$

$0.28m_2 = 0.92 \times 50$

$$m_2 = \frac{46}{0.28}$$

$m_2 = 164kg/h$

Therefore the wet pulp enters at 164kg/h

(b) Let y_3 be the mole fraction of water in the outlet air

Therefore, $y_3P = \dfrac{H_R}{100} \times P_{H_2O}$ (at t = 55°C)

From tables, P_{H_2O} (at 55°C) = 119.6mmHg = 11.96cmHg

Therefore, $y_3 = \dfrac{H_R \times P_{H_2O}}{100 \times P}$

$$= \frac{45 \times 11.96}{100 \times 76}$$

$y_3 = 0.0708$

Let us now take water balance as follows:

Input = Output

Input water = Output in the dried pulp + Output water in air

Hence, $x_2m_2 = x_4m_4 + (y_3m_3 \times$ molecular mass of water)

where m_3 is the outlet air flow rate in kmol/h. Note that y_3m_3 has been multiplied by the molecular mass of water in order to convert it from molar flow rate to mass flow rate since x_2m_2 and x_4m_4 are in mass flow rate.

$x_2m_2 = x_4m_4 + (y_3m_3 \times$ molecular mass of water)

$$0.72 \times 164 = (0.08 \times 50) + (0.0708 \times m_3 \times 18)$$

$$118.08 = 4 + 1.2744m_3$$

$$1.2744m_3 = 118.08 - 4$$

$$1.2744m_3 = 114.08$$

$$m_3 = \frac{114.08}{1.2744}$$

$$m_3 = 89.5\text{kmol/h}$$

The dry air component balance is given by:

Input = Output

$$m_1 = (1 - y_3)m_3$$

where m_1 is the inlet air flow rate in kmols/h

Therefore, $m_1 = (1 - y_3)m_3$

$$m_1 = (1 - 0.0708) \times 89.5$$

$$= 0.9292 \times 89.5$$

$$m_1 = 83.2\text{kmols/h}$$

Let us now convert this molar flow rate to volumetric flow rate as follows:

Recall that: $\quad \dfrac{P_1 V_1 m_2}{T_1} = \dfrac{P_2 V_2 m_1}{T_2} \qquad$ (Refer to example 4)

Therefore, $\quad V_2 = \dfrac{P_1 V_1 m_2 T_2}{P_2 m_1 T_1}$

$$= \frac{76 \times 22.4 \times 83.2 \times (273 + 66)}{76 \times 1 \times 273}$$

$$= \frac{76 \times 22.4 \times 83.2 \times 339}{76 \times 1 \times 273}$$

$$= 2314 \text{m}^3/\text{h}$$

Therefore the flow rate of the inlet air is $2314\text{m}^3/\text{h}$

Note that the inlet air conditions have been used as the final conditions (P_2, V_2, T_2, m_2), while the s.t.p conditions have been used as the initial conditions (P_1, V_1, T_1, m_1).

7. A solution of $MnSO_4.H_2O$ which is saturated at 72°C enters into a recycle stream. The recycled mixture is slightly evaporated and then cooled to 0°C. At 0°C, $MnSO_4.7H_2O$ crystals are formed, and this cooled stream is filtered. The product stream consists of 84% solid crystals and 16% solution. Calculate:

(a). the inlet flow rate that will produce 620kg/h of the crystals

(b). the rate of evaporation of the recycled mixture.

(Take the solubility of $MnSO_4.H_2O$ at 72°C as 61g/100gH_2O, and the solubility of $MnSO_4.7H_2O$ at 0°C as 53g/100gH_2O)

Solution

(a). Note that the feed is at 72°C, while the product is at 0°C. Let us convert the solubility to mass fraction.

Let x_F = mass fraction of $MnSO_4.H_2O$ in the feed

x_P = mass fraction of $MnSO_4.7H_2O$ in the product

Therefore, $x_F = \dfrac{61}{61+100}$

$\quad\quad = 0.3789$

Hence the mass fraction of water in feed = $1 - x_F$

$$= 1 - 0.3789$$

$$= 0.6211$$

Similarly, $x_P = \dfrac{53}{53+100}$

$\quad\quad x_P = 0.3464$

Hence the mass fraction of water in the product = $1 - x_p$

$$= 1 - 0.3464$$

$$= 0.6536$$

Taking overall mass balance on a basis of 620kg/h of $MnSO_4.7H_2O$ produced gives:

Input = Output

$$F = 620 + P + W$$

where F = feed, P = 16% solution and W is the evaporated water

By applying simple proportion, P can be obtained as follows:

84% is equal to 620kg

Therefore, 16% will be equal to: $\frac{16}{84}$ x 620 = 118kg

Therefore, P = 118kg/h solution

Hence, F = 620 + 118 + W

$$F = 738 + W$$

Let us take $MnSO_4.H_2O$ component balance

Input = Output

$$Fx_F = 620(\frac{molecular \quad mass \ of \ MnSO_4.H_2O}{molecular \quad mass \ of \ MnSO_4.7H_2O}) + Px_P$$

Note that molecular mass of $MnSO_4.H_2O$ = 169, while the molecular mass of $MnSO_4.7H_2O$ = 277

Again, $Fx_F = 620(\frac{molecular \quad mass \ of \ MnSO_4.H_2O}{molecular \quad mass \ of \ MnSO_4.7H_2O}) + Px_P$

$$F(0.3789) = 620(\frac{169}{277}) + (118 \ x \ 0.3464)$$

Note that $620(\frac{169}{277})$ gives the mass of $MnSO_4.H_2O$ in the product crystals

$$0.3789F = 378.3 + 40.9$$

$$F = \frac{419.2}{0.3789}$$

$$F = 1106$$

Therefore the inlet feed flow rate is 1106kg/h

(b). Recall that: F = 738 + W

Therefore,   W = 1106 - 738

W = 368

Therefore the rate of evaporation of the recycled mixture is 368kgH$_2$O/h

EXERCISE

1. A distillation column is used to separate 2000kg/h of a mixture containing 68% water and 32% ethanol. The product at the top of the column contains 94% ethanol, while the product at the bottom contains 85% water. The vapour enters the condenser from the top of the column at 1250kg/h and is returned to column as reflux. The rest is withdrawn. Calculate:

(a). the rate of formation of the top product and the bottom product

(b). the flow rate of obtaining ethanol in the top product

(c). the rate of production of ethanol in the bottom product

(d). the ratio of the amount refluxed to the product withdrawn

2. An alcohol-water mixture contains 40% by mass of alcohol. A distillation column is used to separate the mixture by a continuous process. If the composition of water in the top product is 10%, and it is 88% at the bottom, determine:

(a). the amount of alcohol in the top product

(b). what percentage of the feed is the top product

(c). the amount of alcohol in the top product as a percentage of alcohol in the feed

(d) the amount of water in the bottom product.

3. Water and air flow into a humidifier in which the water evaporates completely into the air. The entering air contains 2.2mole% water vapour. The outlet humidified air contains 16mole% water. Calculate the volume of water in m^3/min required to humidify 500mole/min of the entering air. (Take the molar mass of H$_2$O as 18g/mol and the density of H$_2$O as 1g/cm^3)

4. An air conditioner takes in 200m^3/min of air at 28°C, 740mmHg pressure and 85% relative humidity, and gives out the air at 8°C. Calculate:

(a). the flow rate of the outlet air

(b). the rate at which condensed water must be removed from the air conditioner.
(Take the SVP of water at 28°C as 28.3mmHg and at 8°C as 8mmHg)

5. A gas containing vapour enters into a condenser. The partial pressure of the vapour in the gas is 780mmHg. The partial pressure of the vapour in the outlet gas is 740mmHg at a temperature of 110°C. If the total pressure of the system is 850mmHg, calculate the volume of the gas leaving the condenser when 50kmols of the vapour condenses out.

6. Wet paper pulp with a moisture content of 46% enters a continuous process dryer and leaves at a rate of 200kg/h with a moisture content of 12%. Dry air at 62°C and 760mmHg enters the dryer. The air and water vapour leaves the dryer at 48°C and 760mmHg. Calculate:

(a). the rate at which the wet pulp enters the dryer

(b). the flow rate of the inlet air in m^3/h if the outlet air has a relative humidity of 41%
(Take the SVP of water at 48°C as 83.7mmHg)

7. A solution of $MnSO_4.H_2O$ which is saturated at 70°C enters into a recycle stream. The recycled mixture is slightly evaporated and then cooled to 0°C. At 0°C, $MnSO_4.7H_2O$ crystals are formed, and this cooled stream is filtered. The product stream consist of 92% solid crystals and 8% solution. Calculate:

(a). the inlet flow rate that will produce 800kg/h of the crystals

(b). the rate of evaporation of the recycled mixture.

(Take the solubility of $MnSO_4.H_2O$ at 70°C as 58g/100gH$_2$O, and the solubility of $MnSO_4.7H_2O$ at 0°C as 52g/100gH$_2$O)

CHAPTER 5
BALANCES ON SOLVENT EXTRACTION

In simple solvent extraction, a liquid is extracted (separated) from its mixture with another liquid in which both are miscible, i.e. they are soluble in each other. In order to carry out the extraction, a third liquid is used. The liquid to be extracted is soluble in this third liquid, while this third liquid is insoluble in the liquid that will be left unextracted. For example, acetone and hexane mix together to form a homogeneous mixture. When water is added to this mixture, the acetone which is soluble in water dissolves in the water, and hence it is extracted from its mixture with hexane. Water and hexane are insoluble (immiscible). Therefore, water has been used to extract acetone from its mixture with hexane. The water-acetone product is finally withdrawn.

The following examples cover balances on separation processes involving solvent extraction.

Examples

1. A 16% by weight acetone in water enters a single stage extraction unit at a feed rate of 1200litres/h. 92% of the acetone is to be extracted using chloroform at 25°C. Calculate the flow rate of the chloroform into the process.

(Take K = $\dfrac{(x_A)c \text{ phase}}{(x_A)w \text{ phase}}$ = 1.72, where K is the distribution coefficient for the acetone -chloroform - water mixture. $(x_A)c$ is the mass fraction of acetone in chloroform, $(x_A)w$ is the mass fraction of acetone in water. Also, take density of acetone as 0.794g/cm^3 and the density of water as 1.0g/cm^3)

Solution

Let us take a basis of 1200litres/h of feed.

Therefore, the density of the feed is calculated as follows:

$$\frac{1}{\rho_F} = \frac{x_A}{\rho_A} + \frac{x_W}{\rho_W}$$

where ρ represents density, while F, A, and W, represent feed, acetone and water respectively. x_A is the mass fraction of acetone in feed, while x_W is the mass fraction of water in the feed.

Therefore, $\dfrac{1}{\rho_F} = \dfrac{x_A}{\rho_A} + \dfrac{x_W}{\rho_W}$

$$= \frac{0.16}{0.794} + \frac{0.84}{1} \qquad \text{(Note that } x_W = 100 - 16 = 84\% = 0.84)$$

40

$$= \frac{0.16 + 0.667}{0.794}$$

$$\frac{1}{\rho_F} = \frac{0.827}{0.794}$$

Therefore $\quad \rho_F = \dfrac{0.794}{0.827}$

Recall that: Density, $\rho = \dfrac{mass}{volume}$

Hence,   mass = ρ x volume

Therefore, mass flow rate of feed = ρ_F x volumetric flow rate

$$= \frac{0.794}{0.827} \text{ x } 1200$$

$$= 1152 \text{kg/h}$$

Therefore, acetone in feed, F_A = 0.16 x 1152

$$= 184.3 \text{kg/h}$$

Water in feed, F_W = 0.84 x 1152

$$= 967.7 \text{kg/h}$$

Since 92% of the acetone is extracted from the feed, then:

$\quad P_{AE}$ = 0.92 x F_A

$\qquad$ = 0.92 x 184.3

$\qquad$ = 169.6kg/h

where P_{AE} is the acetone extracted by chloroform.

Let us take the acetone component balance:

$\quad$ Input = Output

Hence, $F_A = P_{AL} + P_{AE}$

where P_{AL} is acetone left in the water.

Therefore, 184.3 = P_{AL} + 169.6

$$P_{AL} = 184.3 - 169.6$$

$$P_{AL} = 14.7 \text{kg/h}$$

The mass fraction of acetone in the chloroform is given by:

$$(x_A)_C = \frac{P_{AE}}{P_{AE} + P_C}$$

where P_C is the chloroform in the product.

Hence, $(x_A)_C = \dfrac{169.6}{169.6 + P_C}$

$$= \frac{169.6}{169.6 + F_C}$$

where F_C is the chloroform feed rate. It is equal to P_C since it is assumed that chloroform is insoluble in water. Also, the mass fraction of acetone in water is given by:

$$(x_A)_W = \frac{P_{AL}}{P_{AL} + P_W}$$

where P_W is the water in the product. It is equal to F_W since all the water passes to the product.

Therefore, $(x_A)_W = \dfrac{P_{AL}}{P_{AL} + P_W}$

$$= \frac{14.7}{14.7 + F_W}$$

$$= \frac{14.7}{14.7 + 967.7}$$

$$= \frac{14.7}{982.4}$$

$$(x_A)_W = 0.01496$$

Recall that: $K = \dfrac{(x_A)c \text{ phase}}{(x_A)w \text{ phase}} = 1.72$

Therefore, $\dfrac{\frac{169.6}{169.6 + F_C}}{0.01496} = 1.72$

$$\frac{169.6}{169.6 + F_C} = 0.01496 \times 1.72$$

$$\frac{169.6}{169.6 + F_C} = 0.02573$$

$$169.6 + F_C = \frac{169.6}{0.02573}$$

$$169.6 + F_C = 6591.5$$

$$F_C = 6591.5 - 169.6$$

$$F_C = 6422 kg/h$$

Therefore the flow rate of the chloroform into the process is 6422kg/h

2. A mixture of 72% by weight acetone and 28% by weight hexane is mixed with an equal mass of water. The overall mixture is shaken and allowed to stand. The acetone - water phase is withdrawn. The same amount of water is again added to the mixture left (i.e. hexane phase) and the process is carried out again. What percentage of the acetone in the feed is left unextracted in the hexane?

(Take K = $\frac{(x_A)_H}{(x_A)_W}$ = 0.34, where $(x_A)_H$ is the mass fraction of acetone in hexane, while $(x_A)_W$ is the mass fraction of acetone in water)

Solution

Here it is assumed that water and hexane are immiscible.

Let us take a basis of 100kg feed. This will consists of 72kg (72%) acetone and 28kg (28%) of hexane. The acetone component balance is given by:

$$F_1 = R_1 + E_1$$

where F = feed, R = raffinate, E = extract, and 1 represent stage 1. Note that the raffinate is the unextracted acetone in the hexane, while the extract is the extracted acetone in the water.

Therefore, $F_1 = R_1 + E_1$

$$72 = R_1 + E_1 \text{ ....Equation 1}$$

The mass fraction of acetone in hexane (i.e. the raffinate mass fraction) is given by:

$$(x_A)_H = \frac{R_1}{R_1 + 28} \qquad \text{(Note that hexane = 28kg, and all the hexane remains in the raffinate)}$$

43

Similarly, the extract mass fraction is given by:

$$(x_A)_W = \frac{E_1}{E_1 + 100} \quad \text{(Note that water = 100kg, and all the water remains in the extract)}$$

Note that the mass of water is the same as the original feed mixture which is 100kg, since the feed was mixed with an equal mass of water.

Hence, $\quad \dfrac{(x_A)_H}{(x_A)_W} = 0.34 \quad$ (As given in the question)

$$\frac{\dfrac{R_1}{R_1 + 28}}{\dfrac{E_1}{E_1 + 100}} = 0.34 \quad ....\text{Equation 2}$$

From equation 1, $R_1 = 72 - E_1$ ....Equation 3

Substitute $72 - E_1$ for R_1 in equation 2. This gives:

$$\frac{\dfrac{R_1}{R_1 + 28}}{\dfrac{E_1}{E_1 + 100}} = 0.34 \quad ....\text{Equation 2}$$

$$\frac{\dfrac{72 - E_1}{72 - E_1 + 28}}{\dfrac{E_1}{E_1 + 100}} = 0.34$$

$$\frac{72 - E_1}{100 - E_1} \times \frac{E_1 + 100}{E_1} = 0.34$$

$$0.34E_1(100 - E_1) = (72 - E_1)(E_1 + 100)$$

$$34E_1 - 0.34E_1^2 = 72E_1 + 7200 - E_1^2 - 100E_1$$

$$E_1^2 - 0.34E_1^2 + 34E_1 - 72E_1 + 100E_1 - 7200 = 0$$

$$0.66E_1^2 + 62E_1 - 7200 = 0 \quad \text{(Quadratic equation)}$$

Using the quadratic equation formula to solve this equation gives:

$$E_1 = \frac{-b \pm \sqrt{b^2 - 4ac}}{2a}$$

$a = 0.66, b = 62, c = -7200$

$$E_1 = \frac{-62 \pm \sqrt{62^2 - [4 \times 0.66 \times (-7200)]}}{2 \times 0.66}$$

$$= \frac{-62 \pm \sqrt{3844 + 19008}}{1.32}$$

$$= \frac{-62 \pm \sqrt{22852}}{1.32}$$

$$= \frac{-62 \pm 151}{1.32}$$

$$= \frac{89}{1.32} \qquad \text{(The second answer is discarded since } E_1 \text{ cannot be negative)}$$

E_1 = 67.4kg

From equation 3:

R_1 = 72 - E_1

= 72 - 67.4

R_1 = 4.6kg

This raffinate, R_1 becomes the new feed for the second stage.

Taking acetone component balance in the second stage gives:

$R_1 = R_2 + E_2$

Hence, 4.6 = $R_2 + E_2$ ....Equation 4

In this second stage, the mass fraction of acetone in hexane phase is given by:

$(x_A)_H = \dfrac{R_2}{R_2 + 28}$ (All the hexane (28kg) will continue to be mixed with the raffinate)

The mass fraction of acetone in water phase is given by:

$(x_A)_W = \dfrac{E_2}{E_2 + 100}$ (Note that the same mass (100kg) of water was used in stage 2. All the water will always mix with the extracted acetone)

Therefore, similar to stage 1, the equilibrium in stage 2 is given by:

$$\frac{\dfrac{R_2}{R_2 + 28}}{\dfrac{E_2}{E_2 + 100}} = 0.34 \text{ ....Equation 5}$$

From equation 4, R_2 = 4.6 - E_2 ....Equation 6

Substitute 4.6 - E_2 for R_2 in equation 5. This gives:

$$\frac{\frac{R_2}{R_2+28}}{\frac{E_2}{E_2+100}} = 0.34 \quad\text{......................Equation 5}$$

$$\frac{\frac{4.6-E_2}{4.6-E_2+28}}{\frac{E_2}{E_2+100}} = 0.34$$

$$\frac{4.6-E_2}{32.6-E_2} \times \frac{E_2+100}{E_2} = 0.34$$

$$0.34E_2(32.6 - E_2) = (4.6 - E_2)(E_2 + 100)$$

$$11.08E_2 - 0.34E_2^2 = 4.6E_2 + 460 - E_2^2 - 100E_2$$

$$E_2^2 - 0.34E_2^2 + 11.08E_2 + 100E_2 - 4.6E_2 - 460 = 0$$

$$0.66E_2^2 + 106.48E_2 - 460 = 0$$

Using quadratic equation formula to find E_2 gives:

$$E_2 = \frac{-b \pm \sqrt{b^2 - 4ac}}{2a}$$

a = 0.66, b = 106.48, c = -460

$$E_2 = \frac{-106.48 \pm \sqrt{106.48^2 - [4 \times 0.66 \times (-460)]}}{2 \times 0.66}$$

$$E_2 = \frac{-106.48 \pm \sqrt{11338 + 1214}}{1.32}$$

$$= \frac{-106.48 \pm \sqrt{12552}}{1.32}$$

$$= \frac{-106.48 \pm 112}{1.32}$$

$$= \frac{5.52}{1.32} \quad \text{(The second answer is negative, so it cannot be our answer)}$$

Therefore, E_2 = 4.2kg

From equation 6, R_2 = 4.6 - E_2

$$= 4.6 - 4.2$$

$$R_2 = 0.4kg$$

This is the unextracted acetone in the hexane.

% of acetone in the feed that is left unextracted in hexane is given by:

$$\frac{0.4}{72} \times 100 \quad \text{(Note that the acetone in feed = 72kg)}$$

$$= 0.56\%$$

3. A 24% by weight ethanol in benzene enters a single stage extraction unit at a feed rate of 500kg/h. 86% of the ethanol is to be extracted using water. Calculate the flow rate of the water into the process.

(Take K = $\dfrac{(x_E)_W \text{ phase}}{(x_E)_B \text{ phase}}$ = 1.84, where $(x_E)_W$ is the mass fraction of ethanol in water, $(x_E)_B$ is the mass fraction of ethanol in benzene.

Solution

Ethanol in feed, F_E = 0.24 x 500

$$= 120kg/h$$

Benzene in feed, F_B = 0.76 x 500   (Note that 100 - 24 = 76% benzene in feed)

$$= 380kg/h$$

Since 86% of the ethanol is extracted from the feed, then:

$$P_{EE} = 0.86 \times F_E$$

$$= 0.86 \times 120$$

$$= 103.2kg/h$$

where P_{EE} is the ethanol extracted by water.

Let us take the ethanol component balance:

Input = Output

Hence, $F_E = P_{EL} + P_{EE}$

where P_{EL} is ethanol left in the benzene.

Therefore, $120 = P_{EL} + 103.2$

$$P_{EL} = 120 - 103.2$$

$$P_{EL} = 16.8 kg/h$$

The mass fraction of ethanol in the water is given by:

$$(x_E)_W = \frac{P_{EE}}{P_{EE} + P_W}$$

where P_W is the water in the product.

Hence, $(x_E)_W = \dfrac{103.2}{103.2 + P_W}$

$$= \frac{103.2}{103.2 + F_W}$$

where F_W is the water feed rate. It is equal to P_W since it is assumed that water is insoluble in benzene. Also, the mass fraction of ethanol in benzene is given by:

$$(x_E)_B = \frac{P_{EL}}{P_{EL} + P_B}$$

where P_B is the benzene in the product. It is equal to F_B since all the benzene passes to the product.

Therefore, $(x_E)_B = \dfrac{P_{EL}}{P_{EL} + P_B}$

$$= \frac{16.8}{16.8 + F_B}$$

$$= \frac{16.8}{16.8 + 380}$$

$$= \frac{16.8}{396.8}$$

$$(x_E)_B = 0.04234$$

Recall that: $K = \dfrac{(x_E)_W}{(x_E)_B} = 1.84$

Therefore, $\dfrac{\dfrac{103.2}{103.2+F_W}}{0.04234} = 1.84$

$$\frac{103.2}{103.2+F_W} = 0.04234 \times 1.84$$

$$\frac{103.2}{103.2+F_W} = 0.07791$$

$$103.2 + F_W = \frac{103.2}{0.07791}$$

$$103.2 + F_W = 1324.6$$

$$F_W = 1324.6 - 103.2$$

$$F_W = 1221 kg/h$$

Therefore the flow rate of the water into the process is 1221kg/h

4. A 200kg mixture of 64% by weight acetone and 36% by weight pentane is mixed with 100kg of water. The overall mixture is shaken and allowed to stand. The acetone - water phase is withdrawn. 50kg of water is again added to the mixture left (i.e. pentane phase) and the process is carried out again. What is the total amount of acetone extracted?

(Take $K = \dfrac{(x_A)_p}{(x_A)_w} = 0.42$, where $(x_A)_p$ is the mass fraction of acetone in pentane, while $(x_A)_w$ is the mass fraction of acetone in water)

Solution

Here it is assumed that water and pentane are immiscible.

Mass of acetone in the feed = 0.64 x 200

$$= 128kg$$

Mass of pentane in the feed = 0.36 x 200

$$= 72kg$$

The acetone component balance is given by:

$$F_1 = R_1 + E_1$$

where F = feed, R = raffinate, E = extract, and 1 represent stage 1.

Therefore, $F_1 = R_1 + E_1$

$$128 = R_1 + E_1 \text{ ....Equation 1}$$

The mass fraction of acetone in pentane (i.e. the raffinate mass fraction) is given by:

$$(x_A)p = \frac{R_1}{R_1 + 72} \quad \text{(Pentane = 72kg. This is always mixed with the raffinate)}$$

Similarly, the extract mass fraction is given by:

$$(x_A)w = \frac{E_1}{E_1 + 100} \quad \text{(Water = 100kg. All the water is always mixed with the extract)}$$

Hence, $\dfrac{(x_A)p}{(x_A)w} = 0.42$ (As given in the question)

$$\frac{\frac{R_1}{R_1 + 72}}{\frac{E_1}{E_1 + 100}} = 0.42 \text{ ....Equation 2}$$

From equation 1, $R_1 = 128 - E_1$ ....Equation 3

Substitute $128 - E_1$ for R_1 in equation 2. This gives:

$$\frac{\frac{R_1}{R_1 + 72}}{\frac{E_1}{E_1 + 100}} = 0.42 \text{ ....Equation 2}$$

$$\frac{\frac{128 - E_1}{128 - E_1 + 72}}{\frac{E_1}{E_1 + 100}} = 0.42$$

$$\frac{128 - E_1}{200 - E_1} \times \frac{E_1 + 100}{E_1} = 0.42$$

$0.42E_1(200 - E_1) = (128 - E_1)(E_1 + 100)$

$84E_1 - 0.42E_1^2 = 128E_1 + 12800 - E_1^2 - 100E_1$

$E_1^2 - 0.42E_1^2 + 84E_1 - 128E_1 + 100E_1 - 12800 = 0$

$0.58E_1^2 + 56E_1 - 12800 = 0$ (Quadratic equation)

Using the quadratic equation formula to solve this equation gives:

$$E_1 = \frac{-56 \pm \sqrt{56^2 - [4 \times 0.58 \times (-12800)]}}{2 \times 0.58}$$

$$= \frac{-56 \pm \sqrt{3136 + 29696}}{1.16}$$

$$= \frac{-56 \pm \sqrt{32832}}{1.16}$$

$$= \frac{-56 \pm 181.2}{1.16}$$

$$= \frac{125.2}{1.16} \qquad \text{(The second answer is discarded since } E_1 \text{ cannot be negative)}$$

E_1 = 108kg

From equation 3:

$R_1 = 128 - E_1$

$= 128 - 108$

R_1 = 20kg

This raffinate, R_1, becomes the new feed for the second stage.

Taking acetone component balance in the second stage gives:

$R_1 = R_2 + E_2$

Hence, 20 = $R_2 + E_2$ ....Equation 4

In this second stage, the mass fraction of acetone in pentane phase is given by:

$(x_A)p = \dfrac{R_2}{R_2 + 72}$ (All the pentane (72kg) will continue to be mixed with the raffinate)

The mass fraction of acetone in water phase is given by:

$(x_A)w = \dfrac{E_2}{E_2 + 50}$ (Note that the mass of water that was used in stage 2 is 50kg. All the water will always mix with the extracted acetone)

Therefore, similar to stage 1, the equilibrium in stage 2 is given by:

51

$$\frac{\dfrac{R_2}{R_2+72}}{\dfrac{E_2}{E_2+50}} = 0.42 \quad \text{.....................Equation 5}$$

From equation 4, $R_2 = 20 - E_2$ ....Equation 6

Substitute $20 - E_2$ for R_2 in equation 5. This gives:

$$\frac{\dfrac{R_2}{R_2+72}}{\dfrac{E_2}{E_2+50}} = 0.42 \quad \text{.....................Equation 5}$$

$$\frac{\dfrac{20-E_2}{20-E_2+72}}{\dfrac{E_2}{E_2+50}} = 0.42$$

$$\frac{20-E_2}{92-E_2} \times \frac{E_2+50}{E_2} = 0.42$$

$$0.42E_2(92 - E_2) = (20 - E_2)(E_2 + 50)$$

$$38.64E_2 - 0.42E_2^2 = 20E_2 + 1000 - E_2^2 - 50E_2$$

$$E_2^2 - 0.42E_2^2 + 38.64E_2 + 50E_2 - 20E_2 - 1000 = 0$$

$$0.58E_2^2 + 68.64E_2 - 1000 = 0$$

Using quadratic equation formula to find E_2 gives:

$$E_2 = \frac{-68.64 \pm \sqrt{68.64^2 - [4 \times 0.58 \times (-1000)]}}{2 \times 0.58}$$

$$= \frac{-68.64 \pm \sqrt{4711 + 2320}}{1.16}$$

$$= \frac{-68.64 + 83.9}{1.16} \qquad \text{(The second answer is negative, so it cannot be our answer)}$$

Therefore, $E_2 = 13.2$kg

Hence the total amount of acetone extracted $= E_1 + E_2$

$$= 108 + 13.2$$

$$= 121.2 \text{kg}$$

EXERCISE

1. A 22% by weight acetone in water enters a single stage extraction unit at a feed rate of 1500litres/h. 95% of the acetone is to be extracted using chloroform at 25°C. Calculate the flow rate of the chloroform into the process.

(Take K = $\frac{(x_A)_C}{(x_A)_W}$ = 1.75, where K is the distribution coefficient for the acetone -chloroform - water mixture. $(x_A)_C$ is the mass fraction of acetone in chloroform, $(x_A)_W$ is the mass fraction of acetone in water. Also, take density of acetone as 0.788g/cm^3 and the density of water as 1.0g/cm^3)

2. A mixture of 80% by weight acetone and 20% by weight hexane is mixed with an equal mass of water. The overall mixture is shaken and allowed to stand. The acetone - water phase is withdrawn. The same amount of water is again added to the mixture left (i.e. hexane phase) and the process is carried out again. What percentage of the acetone in the feed is left unextracted in the hexane.

(Take K = $\frac{(x_A)_H}{(x_A)_W}$ = 0.32, where $(x_A)_H$ is the mass fraction of acetone in hexane, while $(x_A)_W$ is the mass fraction of acetone in water)

3. A 35% by weight ethanol in benzene enters a single stage extraction unit at a feed rate of 2000kg/h. 82% of the ethanol is to be extracted using water. Calculate the flow rate of the water into the process.

(Take K = $\frac{(x_E)_W}{(x_E)_B}$ = 1.68, where $(x_E)_W$ is the mass fraction of ethanol in water, $(x_E)_B$ is the mass fraction of ethanol in benzene.

4. A 600kg mixture of 55% by weight acetone and 45% by weight pentane is mixed with 400kg of water. The overall mixture is shaken and allowed to stand. The acetone - water phase is withdrawn. 100kg of water is again added to the mixture left (i.e. pentane phase) and the process is carried out again. What is the total amount of acetone extracted.

(Take K = $\frac{(x_A)_P}{(x_A)_W}$ = 0.48, where $(x_A)_P$ is the mass fraction of acetone in pentane, while $(x_A)_W$ is the mass fraction of acetone in water)

5. A 1000kg mixture of 50% by weight acetone and 50% by weight pentane is mixed with 500kg of water. The overall mixture is shaken and allowed to stand. The acetone - water phase is withdrawn. 300kg of water is again added to the mixture left (i.e. pentane phase) and the process is carried out again. What is the total amount of acetone extracted?

(Take K = $\dfrac{(x_A)_P}{(x_A)_W}$ = 0.44, where $(x_A)_P$ is the mass fraction of acetone in pentane, while $(x_A)_W$ is the mass fraction of acetone in water)

CHAPTER 6
PRESSURE IN LIQUID

Pressure is defined as force per unit area. Its S.I unit is N/m^2 which is called Pascal (Pa). Other popular units of pressure are "mmHg" and "atm". Pressure is given by:

$$P = \frac{F}{A}$$

The pressure exerted by the air above us is called atmospheric pressure. Its value is $1.013 \times 10^5 N/m^2$, 760mmHg or 1atm.

In engineering calculations, absolute pressure is normally used. It is given by:

$$P_{absolute} = P_{atmospheric} + P_{gauge}$$

where P_{gauge} is the gauge pressure

In a liquid, the pressure at a level in the liquid is given by:

$$P = \rho g h$$

where ρ is the density of the liquid, h is the depth of level above the liquid surface, while g is the acceleration due to gravity.

Also recall that: density, $\rho = \dfrac{mass}{volume}$ or $\rho = \dfrac{m}{V}$

U - Tube Manometer

This is a device used for measuring gas pressure or a differential pressure. When the two limbs of the manometer are open, atmospheric pressure acts on them. If the difference in height between the liquid levels in both limbs is h, then the pressure being measured is a measure of the gauge pressure. However, if one limb of the tube is sealed, then the h will give a measure of the absolute pressure.

The sealed tube can function as a barometer when the pressure at the open limb is atmospheric.

Manometric Pressure Difference

When a manometer open at both ends is connected across an orifice plate in a pipeline through which a fluid of density ρ is flowing, then the pressure drop or pressure difference across the orifice plate is given by:

$$P_1 - P_2 = (\rho_F - \rho)gh_F$$

Or $\Delta P = (\rho_F - \rho)gh_F$

where P_1 is the upstream pressure before the orifice plate, P_2 is the downstream pressure after the orifice plate, ΔP is the pressure difference/pressure drop, ρ_F is the density of the liquid in the manometer, and h_F is the difference in height between the liquid levels in the two limbs of the manometer.

If the fluid whose pressure is being measured is a gas, then $(\rho_F - \rho)$ becomes approximately ρ_F since ρ (i.e. the gas density) is very small. Hence the equation above becomes:

$\Delta P = \rho_F g h_F$

The Inclined Manometer

For an inclined manometer, h_F in the equation of a vertical manometer has to be replaced by the expression:

$h_F = L\sin\theta$ (Since $\sin\theta = \dfrac{\text{opposite}}{\text{hypotenuse}} = \dfrac{h_F}{L}$, i.e. trigonometric ratio for an inclined plane)

where L is the slant height (hypotenuse) of the manometer, while θ is the angle of inclination of the manometer to the horizontal. Generally, $\sin\theta = 0.1$.

The manometric equation for an inclined manometer is given by:

$P_1 - P_2 = (\rho_F - \rho)gL\sin\theta$ (h_F is replaced by $L\sin\theta$)

For a gas pressure measurement, the equation becomes:

$P_1 - P_2 = \rho_F gL\sin\theta$

Examples

1. A piece of wood in the shape of a cylinder is 2.8m long, and it floats vertically in a water tank. If the part of the wood above the water surface is 1.5m,

(a). calculate the density of the wood.

(b). If the liquid in the tank is one whose density is unknown and the portion of the wood above the liquid surface is 1.9m, calculate the density of the liquid.

(Density of water = 1000kg/m^3)

Solution

(a). **Method 1:**

From Archimedes's principle, when a body floats in a liquid, the mass of the liquid displaced is equal to the mass of the body.

Let the cross-sectional area of the wood be A. Length of wood submerged in the water is:

2.8 - 1.5 = 1.3m

Therefore, volume of water displaced by the wood is:

V = A x 1.3   [Note that volume = cross-sectional area x length (or height)]

V = 1.3A

Recall that: $\rho = \dfrac{m}{V}$

Hence, m = ρV

Therefore mass of water displaced = 1000 x 1.3A

= 1300A

Similarly, mass of wood = ρ_W x V_W (where ρ_W is the density of wood and V_W is the volume of wood)

But, V_W = length of wood x cross sectional area of wood

= 2.8A

Hence, mass of wood = ρ_W x V_W

= ρ_W x 2.8A

= 2.8Aρ_W

Mass of wood = mass of water displaced   (Archimedes's principle)

Therefore,   2.8Aρ_W = 1300A

$$\rho_W = \frac{1300A}{2.8A}$$

ρ_W = 464.3kg/m^3

Method 2

From method 1 above, it can be established that the density of a solid of uniform cross-sectional area floating in a liquid is given by:

$$\text{Density of floating solid} = \frac{\text{Density of liquid x submerged length of solid}}{\text{Total length of solid}}$$

Therefore, density of wood, $\rho_W = \dfrac{1000 \times 1.3}{2.8}$

$$\rho_W = 464.3 \text{kg/m}^3$$

(b). **Method 1**

$$\text{Density of floating solid} = \frac{\text{Density of liquid x submerged length of solid}}{\text{Total length of solid}}$$

$$464.3 = \frac{\rho_L \times (2.8 - 1.9)}{2.8} \qquad (\rho_L = \text{density of liquid})$$

$$464.3 = \frac{0.9\rho_L}{2.8}$$

Therefore, $\qquad \rho_L = \dfrac{464.3 \times 2.8}{0.9}$

$$\rho_L = 1444.5 \text{kg/m}^3$$

Method 2

When a solid float in a liquid, the length of the solid submerged is inversely proportional to the density of the liquid. This is expressed as follows:

$$L \, \alpha \, \frac{1}{\rho}$$

When such a body floats in two different liquids, then the lengths submerged and the densities of the liquids are related by the expression below:

$$\frac{L_A}{L_B} = \frac{\rho_B}{\rho_A}$$

By using water and the new liquid in question (b) above, we have:

$$\frac{L_{Wt}}{L_{NL}} = \frac{\rho_{NL}}{\rho_{Wt}}$$

where NL represents new liquid, Wt represents water, ρ represents density, while L represents length of solid submerged.

Therefore, $\qquad \dfrac{L_{Wt}}{L_{NL}} = \dfrac{\rho_{NL}}{\rho_{Wt}}$

$$\frac{2.8 - 1.5}{2.8 - 1.9} = \frac{\rho_{NL}}{1000}$$

$$\frac{1.3}{0.9} = \frac{\rho_{NL}}{1000}$$

$$\rho_{NL} = \frac{1000 \times 1.3}{0.9}$$

ρ_{NL} = 1444.4kg/m^3 (As obtained before)

2. A storage tank is used to supply water to a vat. A pressure of at least 2.2atm gauge pressure is needed at the inlet of the vat.

(a). Determine the minimum height from the water level in the tank to the inlet of the vat.

(b). If diesel of density 0.84kg/Litre is used, what will be the height required.

Solutions

(a). Recall that: $\Delta P = \rho_F g h_F$

But, ΔP = 2.2atm

Convert this pressure to N/m^2 by multiplying it by 1.013 x 10^{5}N/m^2

Therefore, 2.2atm = (2.2 x 1.013 x 10^5)N/m^2

$$= 2.229 \times 10^5 \text{N/m}^2$$

Therefore, $\Delta P = \rho_F g h_F$

$$h_F = \frac{\Delta P}{\rho_F g}$$

$$= \frac{2.229 \times 10^5}{1000 \times 9.8} \quad \text{(Note that density of water = 1000kg/m}^3 \text{ and g = 9.8m/s}^2\text{)}$$

h_F = 22.7m

(b). Density of diesel = 0.84kg/Litre. Multiply it by 1000 to convert it to density in kg/m^3.

Therefore, density of diesel = 0.84 x 1000

$$= 840 \text{kg/m}^3$$

Therefore, $\Delta P = \rho_F g h_F$

Hence, $h_F = \dfrac{\Delta P}{\rho_F g}$

$$= \dfrac{2.229 \times 10^5}{840 \times 9.8}$$

$h_F = 27.1m$

3. A manometer closed at one end contains liquid of unknown density. The difference between the liquid levels is 5.2m, when the barometer reading is 780mmHg.

(a). Determine the density of the liquid

(b). The same liquid is used in a manometer connected across an orifice plate in a pipeline through which water is flowing. If the difference in levels of the liquid in the manometer is 25cm with the same barometric reading, calculate the pressure drop in mmHg.

(c). What is the pressure in mmHg downstream of the pipe?

Solutions

(a). $P_1 = 780$mmHg. This can be converted to pressure in N/m^2 as follows:

$P_1 = (\dfrac{780}{760} \times 1.013 \times 10^5)$ (Since 760mmHg $= 1.013 \times 10^5 N/m^2$)

$P_1 = 1.04 \times 10^5 N/m^2$

$P_2 = 0$, since one end of the tube is closed.

Hence, $P_1 - P_2 = \rho_F g h_F$

$P_1 = \rho_F g h_F$ (Since $P_2 = 0$)

$\rho_F = \dfrac{P_1}{g h_F}$

$$= \dfrac{1.04 \times 10^5}{9.8 \times 5.2}$$

$\rho_F = 2041$kg/m^3

Therefore the density of the liquid is 2041kg/m^3

(b). $\Delta P = (\rho_F - \rho)g h_F$ $(h_F = \dfrac{25}{100} = 0.25m)$

$$= (2041 - 1000) \times 9.8 \times 0.25$$

$$= 2550 \text{N/m}^2$$

In order to convert this pressure to pressure in mmHg, divide it by 1.013×10^5 and multiply the value by 760. This gives:

$$\frac{2550}{1.013 \times 10^5} \times 760$$

$$= 19.1 \text{mmHg}$$

(c) $\Delta P = P_1 - P_2$ (P_2 is the pressure downstream)

$$P_2 = P_1 - \Delta P$$

$$= 780 - 19.1$$

$$P_2 = 760.9 \text{mmHg}$$

Therefore the pressure downstream the pipe is 760.9mmHg

4. A manometer open at both ends contains three different liquids. The middle liquid y, forms a J shape in the tube such that the difference between its two levels is h_1. The top liquid z on the right arm of the tube makes a height of h_2. The liquid x at the left arm of the tube is on liquid y, and at the shorter arm of the J shape of y. The tops of x and z are at the same level.

(a). Find an equation for this manometer.

(b). Calculate the pressure at the left open arm of the tube given that the pressure on the right is $1.2 \times 10^5 \text{N/m}^2$ and $h_1 = 30\text{cm}$, $h_2 = 40\text{cm}$, $\rho_X = 840\text{kg/m}^3$, $\rho_Y = 1350\text{kg/m}^3$, and $\rho_Z = 1000\text{kg/m}^3$.

Solutions

(a). $\Delta P = (\rho_F - \rho)gh_F$

Since there are three liquids in this manometer, $(\rho_F - \rho)gh_F$ will be computed into two separate values and then added.

With ρ_X as the reference density, (since it is the lowest), we compute $(\rho_F - \rho)gh_F$ for liquid y as follows:

$$(\rho_Y - \rho_X)gh_1$$

Similarly, $(\rho_F - \rho)gh_F$ is computed for liquid z as follows:

$$(\rho_Z - \rho_X)gh_2$$

Adding these two expressions gives the pressure drop and the equation for the manometer as follows:

$$\Delta P = (\rho_Y - \rho_X)gh_1 + (\rho_Z - \rho_X)gh_2$$

(b). $h_1 = 30cm = (\dfrac{30}{100})m = 0.3m$

$h_2 = 40cm = (\dfrac{40}{100})m = 0.4m$

$$\Delta P = (\rho_Y - \rho_X)gh_1 + (\rho_Z - \rho_X)gh_2$$

$$P_1 - P_2 = (\rho_Y - \rho_X)gh_1 + (\rho_Z - \rho_X)gh_2$$

$$= [(1350 - 840) \times 9.8 \times 0.3] + [(1000 - 840) \times 9.8 \times 0.4]$$

$$= 1499.4 + 627.2$$

$$P_1 - P_2 = 2127$$

$$P_1 = 2127 + P_2$$

$$P_1 = 2127 + (1.2 \times 10^5)$$

$$P_1 = 122127 N/m^2$$

5. An inclined manometer of length 15cm is inclined at an angle of 10°. It is used in an orifice plate which is inserted in a pipeline through which a gas is flowing. Calculate the pressure difference across the orifice plate if the manometer contains mercury of density $13600 kg/m^3$.

Solution

For an inclined manometer we have:

$$\Delta P = (\rho_F - \rho)gLsin\theta$$

$\Delta P = \rho_F gLsin\theta$ (Since ρ is very small when compared to ρ_F)

$$= 13600 \times 9.8 \times 0.15 \times sin10 \quad (15cm = 0.15m)$$

$$= 13600 \times 9.8 \times 0.15 \times 0.1736$$

$\Delta P = 3471N/m^2$

This can be converted to pressure in mmHg as follows:

$$\Delta P = \frac{3471}{1.013 \text{ x } 10^5} \text{ x } 760$$

$\Delta P = 26.0mmHg.$

EXERCISE

(Take the value of g as $9.8m/s^2$)

1. A piece of wood in the shape of a cylinder is 1.6m long, and it floats vertically in a water tank. If the part of the wood above the water surface is 0.9m,

(a). calculate the density of the wood.

(b). If the liquid in the tank is one whose density is unknown and the portion of the wood above the liquid surface is 0.4m, calculate the density of the liquid.

(Density of water = $1000kg/m^3$)

2. A storage tank is used to supply water to a vat. A pressure of at least 1.8atm gauge pressure is needed at the inlet of the vat.
(Density of water = $1000kg/m^3$)

(a). Determine the minimum height from the water level in the tank to the inlet of the vat.

(b). If petrol of density 0.78kg/Litre is used, what will be the height required.

3. A manometer closed at one end contains liquid of unknown density. The difference between the liquid levels is 3.5m, when the barometer reading is 740mmHg.

(a). Determine the density of the liquid

(b). The same liquid is used in a manometer connected across an orifice plate in a pipeline through which water is flowing. If the difference in levels of the liquid in the manometer is 40cm with the same barometric reading, calculate the pressure drop in mmHg.

(c). What is the pressure in mmHg downstream of the pipe?

4. A manometer open at both ends contains three different liquids. The middle liquid q, forms a J shape in the tube such that the difference between its two levels is h_1. The top liquid r on the right arm of the tube makes a height of h_2. The liquid p at the left arm of the tube is on liquid q, and at the shorter arm of the J shape of q. The tops of p and r are at the same level.

(a). Find an equation for this manometer.

(b). Calculate the pressure at the left open arm of the tube given that the pressure on the right is $1.6 \times 10^5 \text{N/m}^2$ and $h_1 = 20\text{cm}$, $h_2 = 35\text{cm}$, $\rho_p = 790\text{kg/m}^3$, $\rho_q = 1200\text{kg/m}^3$, and $\rho_r = 950\text{kg/m}^3$.

5. An inclined manometer of length 24cm is inclined at an angle of $14°$. It is used in an orifice plate which is inserted in a pipeline through which a gas is flowing. Calculate the pressure difference across the orifice plate if the manometer contains mercury of density 13600kg/m^3.

CHAPTER 7
HUMIDITY AND WATER VAPOUR IN THE AIR

Humidity is the amount of water vapour present in the air. When air contains the maximum amount of water vapour that it can carry, then the air is said to be saturated at that air temperature and pressure.

At saturation, the expression for Raoult's law applies as follows:

$$p = yP$$

where p is the partial pressure of the saturated vapour, y is the mole fraction of the vapour, while P is the pressure of the moist air.

The expression above can also be expressed as:

$$yP = P_{H_2O}(T)$$

where $P_{H_2O}(T)$ is the vapour pressure of water at the temperature, T, while P is the pressure of the moist air.

Relative Humidity

This is the ratio of the amount of water in the air to the amount of water that will saturate the air at a particular temperature and pressure. It is usually expressed as a percentage. It is expressed mathematically as follows:

$$H_R = \frac{P_{H_2O}}{P_{satd}} \times 100$$

where H_R = relative humidity, P_{H_2O} = partial pressure of water vapour in the air, and P_{satd} = pressure of the saturated air at the prevailing temperature.

Molal Humidity

This is the ratio of the number of moles of water vapour to the number of moles of the vapour free air. It is expressed as:

$$H_m = \frac{\text{Number of moles of water vapour}}{\text{Number of moles of dry air}}$$

Or, $$H_m = \frac{\text{Mole fraction of water vapour}}{\text{Mole fraction of dry air}}$$

It can also be expressed as:

$$H_m = \frac{P_{H_2O}}{P - P_{H_2O}}$$

where P = total pressure, while P_{H_2O} = partial vapour pressure.

Absolute Humidity

This is the amount of water vapour present in a specific amount of air. It is expressed as grams of water vapour per cubic metre of air, i.e. g/m^3. It can also be expressed as grams of water vapour per kilogram of air, i.e. g/kg. Mathematically, it is expressed as:

$$H_A = \frac{Mass\ of\ water\ vapour}{Mass\ of\ dry\ air})$$

Examples

1. Water and air are contacted in a closed pot at a temperature of 70°C and a pressure of 760mmHg. Calculate the molar composition of the air-vapour mixture.

Solution

From steam tables, the saturated vapour pressure of water at 70°C is 236.8mmHg.

Therefore, $yP = P_{H_2O}(T)$

$$y = \frac{P_{H_2O}(70°C)}{P}$$

$$= \frac{236.8}{760}$$

Hence, $y_{H_2O} = 0.3116$

Therefore, the mole fraction of dry air is:

$y_{Air} = 1 - y_{H_2O}$

$= 1 - 0.3116$

$y_{Air} = 0.6884$

2. On a particular day, the temperature of the environment was found to be 28°C at a pressure of 730mmHg and relative humidity of 90%. Calculate:

(a). the mole fraction of water vapour in the air

(b). the molal humidity

(c). the dew point

(d). the absolute humidity

<u>Solutions</u>

(a). Raoult's law indicates that:

$$yP = P_{H_2O} \quad ....Equation\ 1$$

But relative humidity is given by:

$$H_R = \frac{P_{H_2O}}{P_{H_2O}(t^oC)} \times 100$$

Therefore, $\quad P_{H_2O} = \dfrac{H_R P_{H_2O}(t^oC)}{100} ....Equation\ 2$

Note that P_{H_2O} = partial pressure of water vapour, while $P_{H_2O}(t^oC)$ = saturated vapour pressure at t^oC.

Comparing equations 1 and 2 above, shows that they are both equal to P_{H_2O}. Therefore:

$$yP = \frac{H_R P_{H_2O}(t^oC)}{100}$$

Hence, $\quad y = \dfrac{H_R P_{H_2O}(t^oC)}{100 \ x \ P} \quad ....Equation\ 3$

From steam tables, P_{H_2O}(at 28°C) = 28.7mmHg. Substituting known values into equation 3 above gives:

$$y = \frac{H_R P_{H_2O}(t^oC)}{100 \ x \ P} \quad ....Equation\ 3$$

$$y = \frac{90 \ x \ 28.7}{100 \ x \ 730}$$

$$y_{H_2O} = 0.0354$$

(b). Molal humidity is given by:

$$H_m = \text{mole fraction of water vapour/mole fraction of dry air}$$

$$= \frac{0.0354}{1 - 0.0354}$$

$$= \frac{0.0354}{0.9646}$$

$$= 0.0367 \text{mol } H_2O/\text{mol dry air}$$

(c). The dew point is the temperature at which this 0.0354 moles of water will saturate the air.

$$P_{H_2O}(t^\circ C)_{\text{Dew Point}} = y_{H_2O}P$$

$$= 0.0354 \times 730$$

$$P_{H_2O}(t^\circ C)_{\text{Dew Point}} = 25.8$$

From steam tables, the temperature that corresponds to this saturated vapour pressure is 26.2.

Therefore, the dew point, $t_{\text{Dew Point}} = 26.2^\circ C$.

(d). The absolute humidity is given by:

$$H_A = \frac{\text{mass of water vapour}}{\text{mass of dry air}}$$

Recall that: Number of moles $= \dfrac{\text{mass}}{\text{molar mass}}$

Therefore, mass of water vapour = Number of moles x molar mass of water

$$= 0.0354 \times 18 \quad \text{(Note that molecular mass of } H_2O = 18)$$

$$= 0.6372$$

Similarly, mass of dry air = moles of dry air x molar mass of air

$$= (1 - 0.0354) \times 29 \quad \text{(Note that the molecular mass of air = 29)}$$

$$= 0.9646 \times 29$$

$$= 27.9734$$

Therefore, $\quad H_A = \dfrac{\text{mass of water vapour}}{\text{mass of dry air}}$

$$= \frac{0.6372}{27.9734}$$

$$H_A = 0.0228 \text{g } H_2O/\text{g dry air}$$

3. Air of relative humidity 90% enters into a condenser at a temperature of 65°C and a pressure of 750mmHg, and leaves the condenser at a temperature of 5°C. What is the volume of water that condenses out in 1 hour?

Solution

Basis: $1m^3$/min air at 65°C and 750mmHg

Volume of inlet air = $1m^3$/min, (i.e. the volumetric flow rate)

Temperature of inlet air = 65°C

Pressure of inlet air = 750 mmHg

Relative humidity of inlet air = 90%

Temperature of outlet air = 5°C

Also, let:

y_1 = mole fraction of water in inlet air

m_1 = molar flow rate of inlet air in kmol/min

y_2 = mole fraction of water in outlet air

m_2 = molar flow rate of outlet air in kmol/min

m_3 = molar flow rate of condensed water out of the unit in kmol/min

Let us calculate y_1.

Recall that: $y_1 = \dfrac{H_R \times P_{H_2O} \ (\text{at } 65°C)}{100 \times P}$ (As explained in example 2 above)

From tables it can be obtained that:

P_{H_2O} at 65°C = 190 mmHg

Substituting into the equation above gives:

$y_1 = \dfrac{90 \times 190}{100 \times 750}$

$y_1 = 0.228$

Therefore, mole fraction of the dry air in the inlet air = 1 - y_1

1 – 0.228 = 0.772

Let us calculate the molar flow rate of the inlet air. Recall the ideal gas equation:

$$PV = mRT \quad \text{...............Equation 1}$$

At standard conditions, P = 760mmHg, V = 22.4m^3, (molar volume of gas), T = 273K, m = 1kmol (because 1kmol = 22.4m^3)

At the inlet condition we have:

$$P_1V_1 = m_1RT_1 \quad \text{...................Equation (2)}$$

Where P_1 = 750mmHg, V_1 = 1m^3/min, T_1 = 273 + 65 = 338K, m_1 = ?

Dividing equation 1 by equation 2 gives:

$$\frac{PV}{P_1V_1} = \frac{mT}{m_1T_1} \quad \text{(Note that R has cancelled out)}$$

Therefore, $m_1 = \dfrac{P_1V_1mT}{PVT_1}$

$$= \frac{750 \times 1 \times 1 \times 273}{760 \times 22.4 \times 338}$$

$$m_1 = 0.0356 \text{ kmol/min}$$

At the outlet conditions, y_2 can be obtained from Raoult's law as follows:

$$y_2P = P_{H_2O} \text{ (5}^\circ\text{C)}$$

Therefore, $y_2 = \dfrac{P_{H_2O}}{P}$

From tables, P_{H_2O} (at 5°C) = 6.63mmHg

Therefore, $y_2 = \dfrac{6.63}{750}$

$$y_2 = 0.00884$$

Therefore mole fraction of the dry air in the outlet air is:

$$1 - y_2 = 1 - 0.00884 = 0.9912$$

Taking dry air balance:

Input dry air = Output dry air

$(1-y_1)m_1 = (1 - y_2)m_2$

$0.772 \times 0.0356 = 0.9912m_2$

$$m_2 = \frac{0.772 \times 0.0356}{0.9912}$$

$m_2 = 0.0277 \text{kmol/min}$

Taking water balance

Input water = Output water

$m_1y_1 = m_2y_2 + m_3$ (m_3 = molar flow rate of condensed water out of the unit in kmol/min)

$\therefore \quad m_3 = m_1y_1 - m_2y_2$

$\quad = (0.0356 \times 0.228) - (0.0277 \times 0.00884)$

$m_3 = 0.00787 \text{ kmol/min}$

This water obtained can be converted to mass as follows:

$0.00787 \text{kmol} = (0.00787 \times 1000) \text{mols}$

$= 7.87 \text{mols}$

But, Number of moles $= \dfrac{\text{mass}}{\text{Molar mass}}$

$7.87 = \dfrac{\text{mass}}{18}$ (Since H_2O = 18g/mol)

$\therefore \quad \text{Mass} = 7.87 \times 18$

$= 141.7 \text{g}$

Therefore, 141.7g/min of water will condense out of the condense

Water that condensed out in 1 hour

Since in 1 minute, 141.7g of water is condensed out,

Therefore, in 1 hour, water that will condense out is given by:

141.7 x 60   (1 hour = 60 minutes)

= 8502g

= 8.502kg

8.502kg of water will condense in 1 hour

This mass can also be converted to litres to give:

8.502 Litres.   (Because the density of water = 1kg/L)

Therefore, 8.502 Litres of water will condense out of the unit in 1 hour

4. Air of relative humidity 50% enters into a condenser at a temperature of 35°C and a pressure of 680mmHg, and leaves the condenser at a temperature of 5°C.

(a). What is the mass of water that condenses out of the unit in 1min?

(b). What is the volume (in litres) of water that condenses out of the unit in 1 hour?

Solution

Basis: 1m^3/min air at 35°C and 680mmHg

Volume of inlet air = 1m^3/min, (i.e. the volumetric flow rate)

Temperature of inlet air = 35°C

Pressure of inlet air = 680 mmHg

Relative humidity of inlet air be = 50%

Temperature of outlet air = 5°C

Also, let:

y_1 = mole fraction of water in inlet air

m_1 = molar flow rate of inlet air in kmol/min

y_2 = mole fraction of water in outlet air

m_2 = molar flow rate of outlet air in kmol/min

m_3 = molar flow rate of condensed water out of the unit in kmol/min

Let us calculate y_1.

Therefore, $y_1 = \dfrac{H_R \times P_{H_2O} \,(\text{at } 35°C)}{100 \times P}$

From steam tables P_{H_2O} at 35°C = 42.8 mmHg

Substituting into the equation above gives:

$$y_1 = \frac{50 \times 42.8}{100 \times 680}$$

$y_1 = 0.0315$

Therefore, mole fraction of the dry air in the inlet air = $1 - y_1$

$= 1 - 0.0315 = 0.9685$

Let us calculate the molar flow rate of the inlet air. Recall the ideal gas equation:

$PV = mRT$ ....Equation 1

At standard conditions, P = 760mmHg, V = 22.4m^3, (molar volume of gas), T = 273K, m = 1kmol (because 1kmol = 22.4m^3)

At the inlet condition we have:

$P_1V_1 = m_1RT_1$ ....Equation 2

Where P_1 = 680mmHg, V_1 = 1m^3/min, T_1 = 273 + 35 = 308K, m_1 = ?

Dividing equation 1 by equation 2 gives:

$\dfrac{PV}{P_1V_1} = \dfrac{mT}{m_1T_1}$ (Note that R has cancelled out)

Therefore, $m_1 = \dfrac{P_1V_1mT}{PVT_1}$

$= \dfrac{680 \times 1 \times 1 \times 273}{760 \times 22.4 \times 308}$

$m_1 = 0.0354$ kmol/min

At the outlet conditions, y_2 can be obtained from Raoult's law as follows:

$y_2P = P_{H_2O}$

Therefore, $y_2 = \dfrac{P_{H_2O}}{P}$

From tables, P_{H_2O} (at 5°C) = 6.63mmHg

Therefore, $y_2 = \dfrac{6.63}{680}$

$y_2 = 0.00975$

Therefore mole fraction of the dry air in the outlet air is:

$1 - y_2 = 1 - 0.00975$

$= 0.9903$

Taking dry air balance:

Input dry air = Output dry air

$(1-y_1)m_1 = (1 - y_2)m_2$

$0.9685 \times 0.0354 = 0.9903m_2$

$m_2 = \dfrac{0.9685 \times 0.0354}{0.9903}$

$m_2 = 0.0346$ kmol/min

Taking water balance

Input water = Output water

$m_1y_1 = m_2y_2 + m_3$

$\therefore \quad m_3 = m_1y_1 - m_2y_2$

$= (0.0354 \times 0.0315) - (0.0346 \times 0.00975)$

$m_3 = 0.000778$ kmol/min

This water obtained can be converted to mass as follows:

0.000778kmol = (0.000778×1000)mols

$= 0.778$mols/min

But, Number of moles = $\dfrac{\text{mass}}{\text{Molar mass}}$

74

$$0.778 = \frac{\text{mass}}{18} \quad \text{(Since } H_2O = 18\text{g/mol)}$$

$\therefore \quad$ Mass = 0.778 x 18

$= 14.0\text{g/min}$

Therefore, 14.0g of water will condense out of the unit in 1 minute.

(b). Water that condensed out in 1 hour can be obtained as follows.

Since in 1 minute, 14.0g of water is condensed,

Therefore, in 1 hour, water that will condense is given by:

14 x 60   (1 hour = 60 minutes)

= 840g

= 0.84kg

0.84kg of water will condense in 1 hour

This mass can be converted to litres to give:

0.84 Litres.   (Because the density of water = 1kg/L)

Therefore, 0.84 Litres of water will condense out of the unit in 1 hour.

EXERCISE

1. Water and air are contacted in a closed vessel at a temperature of 68°C and a pressure of 760mmHg. Calculate the molar composition of the air-vapour mixture.
(SVP of water at 68°C = 214.2mmHg)

2. On a particular day, the temperature of the environment was found to be 34°C at a pressure of 700mmHg and relative humidity of 60%. Calculate:

(a). the mole fraction of water vapour in the air

(b). the molal humidity

(c). the dew point

(d). the absolute humidity

(SVP of water at 34°C = 39.9mmHg)

3. Air of relative humidity 60% enters into a condenser at a temperature of 45°C and a pressure of 770mmHg, and leaves the condenser at a temperature of 5°C. What is the volume of water that condenses out in 30 minutes?
(SVP of water at 45°C = 71.9mmHg and at 5°C = 6.5mmHg)

4. Air of relative humidity 80% enters into a condenser at a temperature of 40°C and a pressure of 760mmHg, and leaves the condenser at a temperature of 10°C.

(a). What is the mass of water that condenses out of the unit in 30min?

(b). What is the volume (in litres) of water that condenses out of the unit in 1 day?
(SVP of water at 40°C = 55.3mmHg and at 10°C = 9.2mmHg)

5. On a particular day, the temperature of the environment was found to be 22°C at a pressure of 740mmHg and relative humidity of 35%. Calculate:

(a). the mole fraction of water vapour in the air

(b). the molal humidity

(c). the dew point

(d). the absolute humidity

(SVP of water at 22°C = 19.8mmHg)

ANSWERS TO THE EXERCISES

Exercise 1

There is no exercise to chapter 1

Exercise 2

1. 1067.7Kg   2(a) 0.691Kg   (b) Fabric: 90.6%, Moisture: 9.4%   3. 350,541.2Kg/h
4(a) 3.61Kg   (b) 1.39Kg   5. 365.5Kg

Exercise 3

1. 1,428.6kg   2(a) 677.4kg   (b) 1437.1kg   3. $56.27
4. 98% HCl = 13.7litres, 64% HCl = 4.9litres   5. $8.5

Exercise 4

1(a) Top product = 430.4kg/h, Bottom product = 1569.6kg/h   (b) 404.6kg/h
(c) 235.4kg/h   (d) 1.9   2(a) 32.3kg   (b) 35.9%   (c) 80.8%   (d) 56.4kg
3. 1.478×10^{-3} m^3/min   4(a) 182.6m^3/min   (b) 0.173kmols/min   5. 2450.2m^3
6(a) 325.9kg/h   (b) 3960.9m^3/h   7(a) 1394.4kg/h   (b) 524.8kgH_2O/h

Exercise 5

1. 11743kg/h   2. 0.24%   3. 3294kg/h   4. 250kg   5. 453.7kg

Exercise 6

1 (a) 437.5kg/m^3 (b) 583.3kg/m^3 2(a) 18.6m   (b) 23.8m   3(a) 2818kg/m^3
(b) 53.5mmHg   (c) 686.5mmHg   4(a) $\Delta P = (\rho_g - \rho_p)gh_1 + (\rho_r - \rho_p)gh_2$
(b) 161325.4N/m2   5. 58.1mmHg

Exercise 7

1 y_{H_2O} = 0.2818, y_{air} = 0.7182   2(a) 0.0342   (b) 0.0354   (c) 25.1°C   (d) 0.02198
3. 1.0litres   4(a) 982.8g   (b) 47.2litres
5(a) y_{H_2O} = 0.00936   (b) 0.00945   (c) 5.9°C   (d) 0.00587

For other books written by the author, go to amazon and search for the authors name: Kingsley Augustine. You will see all the books written by the author.

For issues, enquiries and suggestions as regards this book, contact the author on:
Email: kingzohb2@yahoo.com

www.ingramcontent.com/pod-product-compliance
Lightning Source LLC
Chambersburg PA
CBHW081614220526
45468CB00010B/2867